W9-CIQ-604

A Journey Along the
Erie Canal

viding Multidigit Numbers by One-Digit Numbers Without Remainders

Janey Levy

PowerMath™

The Rosen Publishing Group's
PowerKids Press™
New York

Published in 2004 by The Rosen Publishing Group, Inc.
29 East 21st Street, New York, NY 10010

Book Design: Michael J. Flynn

Photo Credits: Cover, p. 19 © Lee Snider/Corbis; p. 4 (left inset) © Geoffrey Clements/Corbis; p. 4 (right inset) © Francis G. Mayer/Corbis; pp. 7, 9 © Michael Maslan Historic Photographs/Corbis; p. 11 (both) © Library of Congress; pp. 13, 27 © Bettmann/Corbis; pp. 15, 17 © Corbis; p. 21 © Museum of the City of New York/Corbis; p. 23 © New York State Archives; p. 25 by Michael Flynn; p. 29 © Canal Society of New York State; p. 30 © New York State Canal Commission.

Library of Congress Cataloging-in-Publication Data

Levy, Janey.
  A journey along the Erie Canal : dividing multidigit numbers by
one-digit numbers without remainders / Janey Levy.
       p. cm. — (PowerMath)
Summary: Describes the construction and history of the Erie Canal and
uses the information to illustrate elementary division.
  ISBN 0-8239-8991-7 (lib. bdg.)
  ISBN 0-8239-8904-6 (pbk.)
  6-pack ISBN: 0-8239-7432-4
1. Division—Juvenile literature. 2. Erie Canal (N.Y.)—Juvenile
literature. [1. Division. 2. Erie Canal (N.Y.] I. Title. II. Series.
  QA115.L47 2004
  513.2'14—dc21

                              2002156678

Manufactured in the United States of America

Contents

New York State's Erie Canal

The Erie Canal stretches across New York state from the Hudson River in the east to Lake Erie in the west. During the summer, visitors to the canal can enjoy the beautiful scenery while they take a journey through history on an old-fashioned canal boat.

The canal was completed in 1825 and marked a turning point in U.S. history. It encouraged the settlement of western lands by providing a fast, inexpensive way to get supplies to the settlers. Boats carried supplies from New York City up the Hudson River to the canal, then west on the canal to Buffalo, on the shore of Lake Erie. In return, farm products and lumber from the settlers were hauled back to New York City. The canal soon made New York City the largest and richest city in the nation.

New York City 1816

New York City 1848

▲

Before the Erie Canal was built, it took 3 weeks (21 days) to haul goods from New York City across the state to Buffalo, which is located on Lake Erie. The canal reduced the time required for the journey to 8 days. That means it took 13 fewer days to haul goods from New York City to Lake Erie using the canal!

Building the Erie Canal

No one had ever built anything like the Erie Canal. Before 1817, the longest canal that anyone had built was only 30 miles long. The Erie Canal was more than 12 times that length: it was 363 miles long!

On July 4, 1817, digging for the first section of the canal started at the city of Rome, New York. That section would run about 96 miles from Utica, New York, which is east of Rome, to the Seneca River, which is west of Rome. After about 5 months of work, 15 miles of the canal had been dug. On average, how many miles were dug each month? You can find the answer by dividing 15 miles by 5 months.

$$5 \overline{)\,15\,} \; \begin{array}{r} 3 \\ -15 \\ \hline 0 \end{array}$$

The canal workers dug an average of 3 miles per month.

If the canal workers continued to dig an average of 3 miles each month, how long would it have taken to dig the entire canal? Look at the top of page 7 to find out.

$$\begin{array}{r} 121 \\ 3\overline{)363} \\ \underline{-3} \\ 06 \\ \underline{-6} \\ 03 \\ \underline{-3} \\ 0 \end{array}$$

To find the answer, divide 363 miles by 3 miles per month. The answer is 121 months, or 10 years and 1 month! In fact, that's how long the people in charge of the canal expected it to take.

Many of the laborers who worked on the Erie Canal were Irish immigrants who had come to the United States hoping for a better life.

Building the canal was very hard work. Before the digging could start, trees had to be cut down, and the tree stumps had to be pulled out of the ground. The **engineers** in charge of building the canal invented machines to help the workers do this.

However, there were no machines to help with the digging. The laborers had to do all the work themselves. They used only simple tools like shovels for digging. In the winter, when their shovels could not pierce the frozen ground, they broke it up with sharp tools called pickaxes.

The men who built the canal worked 6 days a week for about 10 hours each day. A full week's work was 60 hours—sometimes more. For 60 hours of work, each worker earned about $3.00 a week. How much did a laborer earn each day? You can divide $3.00 by 6 days to find the answer.

$$
\begin{array}{r}
\$0.50 \\
6\overline{)\$3.00} \\
-3\ 0 \\
\hline
00
\end{array}
$$

A canal worker earned $0.50 each day. This was an average wage during this time!

Much of the canal was built using simple tools like the ones shown in this photograph. The man in the middle is holding a pickax.

The engineers in charge of the Erie Canal earned much more money than the laborers. The chief engineer for the canal was a **surveyor** named Benjamin Wright, who was known for his honesty and exactness. As chief engineer, Wright earned about $36.00 per week. If he worked 6 days a week like the laborers did, how much did he earn in a day? You can divide $36.00 by 6 days to find the answer.

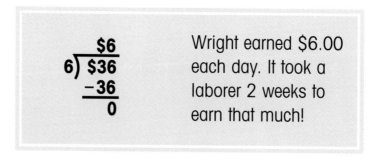

Wright earned $6.00 each day. It took a laborer 2 weeks to earn that much!

Benjamin Wright's assistant was a surveyor named James Geddes. Geddes earned $24.00 for working 6 days a week. How much did he earn in a day? To find the answer, you can divide $24.00 by 6.

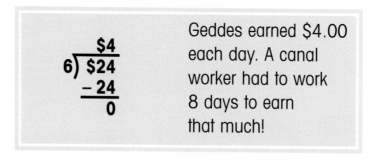

Geddes earned $4.00 each day. A canal worker had to work 8 days to earn that much!

After he finished the Erie Canal, Wright helped to build many other canals. He also helped to build some of the first railroads in the country.

Benjamin Wright

James Geddes

Like Wright, Geddes was known for his honesty and exactness.

11

The engineers had many problems to solve in building the canal. One of the problems was how to get across the rivers and creeks that lay in the path of the canal. To solve this problem, they built a special type of bridge called an **aqueduct**. Altogether, the engineers built 18 aqueduct bridges on the Erie Canal.

At the western end of the canal, near Lake Erie, the engineers faced another **challenge**. To get canal boats to Lake Erie, the engineers had to find a way to get the boats to the top of a 70-foot cliff that was in the path of the canal.

The usual method for moving a boat between sections of a canal that had different water levels was to build a lock. A lock is a part of the canal that has gates at each end. To reach a higher level, a boat on the lower level enters the lock. Then the gates at both ends of the lock are closed. Water is let into the lock until the boat reaches the level of the higher section. The gates at that end of the lock open, and the boat passes through.

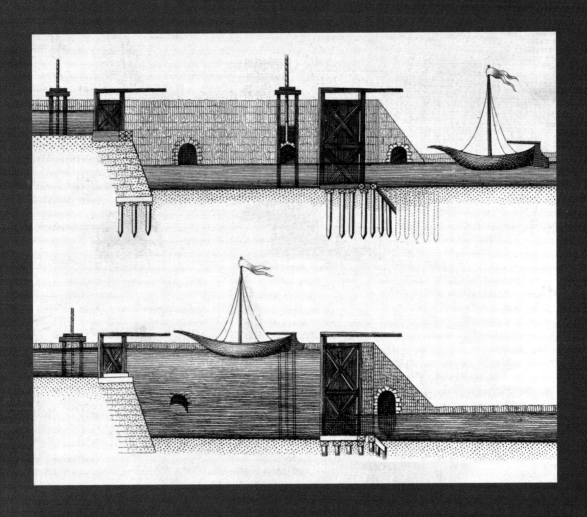

These pictures show how a boat moves through
a lock from a lower level to a higher level.
The process is reversed for boats moving from a
higher level to a lower one.

The part of the canal at the cliff near Lake Erie was very difficult to plan and design. It was not possible to raise a boat 70 feet with just a single lock. Nathan Roberts, an assistant engineer for the Erie Canal, was put in charge of solving the problem. Roberts made a plan for a series of 5 locks that looked like a giant flight of stairs. Nothing this difficult had ever been done before. When Roberts's locks were finished, they became one of the great wonders of the Erie Canal. People all over the United States—and in other countries as well—wanted pictures of them.

Roberts's plan actually called for 2 sets of locks side by side. The locks on one side carried boats moving from the lower level to the higher level. The locks next to them carried boats moving from the higher level to the lower one.

How many feet did each of Roberts's locks raise or lower a canal boat? You can find the answer by dividing 70 feet by 5 locks.

$$
\begin{array}{r}
14 \\
5 \overline{)\,70} \\
-5 \\
\hline
20 \\
-20 \\
\hline
0
\end{array}
$$

Each of Roberts's locks raised or lowered a boat 14 feet.

▲

Artists made drawings and paintings of Roberts's locks. They also took photographs of them. Pictures of the locks even appeared in books published in Paris and London.

When construction of the locks began in 1821, there were only a few log cabins in the area. The tiny community had only 1 street. By 1825, there was a town of 2,500 people. About half of these people were Irish laborers who had come to build the locks. There were also merchants, farmers, doctors, and bankers. The town was named Lockport in honor of the locks that made it famous.

If the weekly pay for all the laborers working on the locks was $3,600.00, and each worker earned $3.00 a week, how many workers were there? You can find the answer by dividing $3,600.00 by $3.00.

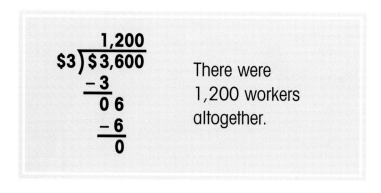

$$\begin{array}{r} 1{,}200 \\ \$3\overline{)\,\$3{,}600} \\ -3 \\ \hline 0\,6 \\ -6 \\ \hline 0 \end{array}$$

There were 1,200 workers altogether.

The engineers created special machines to
lift out the rock that the laborers cut from
the cliff while building the locks.

Travel and Trade on the Erie Canal

Even though laborers were still working on some sections of the canal, the sections that had already been completed were opened for business. On October 23, 1819, a special **ceremony** was held to mark the official opening of the first section of the canal, which ran between Utica and Rome. A canal boat named the *Chief Engineer* set out from Utica and arrived in Rome about 4 hours later. The total distance traveled by the boat was about 16 miles. How many miles per hour did the boat travel? You can divide 16 miles by 4 hours to find the answer.

$$
\begin{array}{r}
4 \\
4 {\overline{)}\, 16}\\
-16\\
\hline
0
\end{array}
$$

The boat traveled 4 miles per hour.

Canal boats moved slowly because they did not have engines or sails. Instead, they were pulled by teams of horses or mules walking on a path beside the canal.

▶

By the end of 1819, the 96-mile section from Utica to the Seneca River was opened. How many hours would it have taken a boat going at the same speed as the *Chief Engineer* to travel the 96 miles from Utica to the Seneca River? To find the answer, divide 96 miles by 4 miles per hour.

$$
\begin{array}{r}
24 \\
4{\overline{\smash{\big)}\,96}} \\
\underline{-8} \\
16 \\
\underline{-16} \\
0
\end{array}
$$

It would take 24 hours to travel from Utica to the Seneca River.

The entire canal was completed in 1825, about 2 years sooner than expected. Immediately the cost for moving goods between Buffalo and New York City dropped sharply. Before the canal was completed, it cost $100.00 per ton to haul wheat from Buffalo to New York City by wagon. On the canal, it cost only $5.00 per ton. How many tons of wheat could a farmer ship on the canal for $100.00? To find the answer, divide $100.00 by $5.00.

$$\begin{array}{r} 20 \\ \$5\overline{)\$100} \\ -10 \\ \hline 00 \end{array}$$

A farmer could ship 20 tons of wheat on the Erie Canal for $100.

If a farmer paid $750.00 to ship wheat from Buffalo to New York City on canal boats, how many tons of wheat did the farmer ship? You can find the answer by dividing $750.00 by $5.00.

$$\begin{array}{r} 150 \\ \$5\overline{)\$750} \\ -5 \\ \hline 25 \\ -25 \\ \hline 00 \end{array}$$

The farmer shipped 150 tons of wheat.

▲

When the canal was completed, the governor of New York rode a
boat called the *Seneca Chief* from Buffalo to New York City. Nearly
150 boats greeted the *Seneca Chief* when it reached New York
Harbor. More than 100,000 people crowded the shores for the
ceremony honoring the event. The painting on this page shows the
boats waiting for the *Seneca Chief* to arrive in New York Harbor.

Part of the cost of shipping goods on the canal came from **tolls** that were paid to New York State. New York State earned money by charging tolls for the freight carried on the canal boats. During the first 9 months the canal was open, the state collected about $756,000.00 in tolls. On average, how much did the state collect each month? To find the answer, divide $756,000.00 by 9 months.

$$
\begin{array}{r}
\$84{,}000 \\
9\,)\overline{\$756{,}000} \\
-\,72 \\
\hline
36 \\
-\,36 \\
\hline
0
\end{array}
$$

The state collected $84,000 in tolls each month!

The canal had cost about $7 million to build. In less than 10 years, the canal had earned more than that in tolls!

This report from an Erie Canal toll collector covers 1 week in 1860. Among the goods listed are 1,040,860 bushels of wheat; 1,087,400 pounds of butter; 1,659,000 pounds of cheese; and 15,900 pounds of wool. ▶

WEEKLY STATEMENT

Showing the quantity of the several articles First Cleared on the Canals at, and the quantity
Left at West Troy for W.Y. during the 4th week in Oct 1860

MERCHANDISE CLEARED.

ARTICLES.	On Erie Canal.	On Champlain Canal.	TOTAL.
Sugar at 2 mills,..................pounds,			
Molasses, " .... "			
Coffee, " .... "			
Nails, " .... "			
Iron, " .... "			
Railroad Iron, " .... "			
All other merchandise at 2 mills,........... "			
TOTAL, ...			

	Left from Erie Canal.	Left from Champlain Canal.	TOTAL.
Flour, ...barrels,	33,114		33,114
Wheat, ...bushels,	1,040,860		1,040,860
Corn, ... "	131,583		131,583
Barley, ... "	150,607		150,607
Rye, ... "	18,113		18,113
Oats, ... "	639,976	3,000	642,976
Bran and Ship Stuffs, ...pounds,	1,200,600		1,200,600
Ashes, ...barrels,	32		32
Beef, ... "	88		88
Pork, ... "			
Bacon, ...pounds,			
Butter, ... "	1,087,400	300	1,087,700
Lard, ... "	40,100		40,100
Cheese, ... "	1,659,000	1,500	1,659,000
Wool, ... "	15,900		15,900
Domestic Spirits, ...gallons,	30,585		30,585

I CERTIFY the above to be correct,

W. L. Sunderlin

Collector.

23

A farmer who shipped 150 tons of wheat from Buffalo to New York City needed 5 canal boats to carry the wheat. How many tons could each boat carry? You can find the answer by dividing 150 tons by 5 boats.

$$\begin{array}{r} 30 \\ 5\overline{)150} \\ -15 \\ \hline 00 \end{array}$$

Each canal boat could carry 30 tons.

By 1830, larger canal boats were being used. The farmer would have needed only 2 canal boats to carry his 150 tons of wheat. How many tons could each boat carry? You can divide 150 tons by 2 boats to find the answer.

$$\begin{array}{r} 75 \\ 2\overline{)150} \\ -14 \\ \hline 10 \\ -10 \\ \hline 0 \end{array}$$

Each canal boat could carry 75 tons.

After the canal was made wider and deeper in 1862, it could handle boats that carried 240 tons. The farmer who needed 5 boats to carry 150 tons of wheat in 1825 could get all his wheat on 1 boat in 1862, with room left for 90 more tons!

This drawing, based on one done around 1900, shows the different sizes of canal boats used on the Erie Canal at different times. It also tells how much each boat could carry.

▼

1817–1830
First boats: 61 feet long, 7 feet wide

30 tons

1830–1850
Later boats on original canal: 75 feet long, 12 feet wide

75 tons

1850–1862
Largest boats on original canal: 90 feet long, 15 feet wide

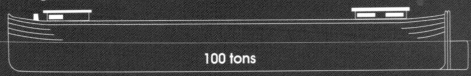

100 tons

1862–1899
Boats used after 1862 enlargement: 98 feet long, $17\frac{1}{2}$ feet wide

240 tons

Special boats called **packet boats** carried **passengers** on the Erie Canal. Inside a packet boat's **cabin** there were benches, as well as a long table on which meals were served. In good weather, passengers could sit in chairs on a deck on top of the cabin's roof. Riding on the deck could be dangerous. Some bridges were so low that passengers had to jump from their chairs and lie flat on the deck while the boat passed under these bridges! This was such a familiar part of a journey on the Erie Canal that it found its way into a popular song written in 1905. The song was called the "Erie Canal Song" or "Low Bridge, Everybody Down."

Passengers paid 4¢ per mile to ride the packet boats. If a passenger paid $3.84—or 384¢—for their trip on a packet boat, how many miles did they travel? Divide 384 by 4 to find the answer.

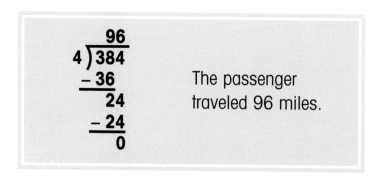

$$
\begin{array}{r}
96 \\
4\overline{)384} \\
-36 \\
\hline
24 \\
-24 \\
\hline
0
\end{array}
$$

The passenger traveled 96 miles.

This picture of a boat approaching a low bridge
on the Erie Canal was painted around 1830.

Railroads started to become a popular way to travel in the 1830s. People still enjoyed the relaxing travel offered by packet boats, but train travel was faster. It became popular to take trips that combined travel by packet boat and travel by train. In the 1840s, one very popular journey was to go from Syracuse to Lockport by packet boat, then from Lockport to Niagara Falls by train. This trip cost about 4¢ per mile, which was the same it would have cost to make the entire journey by packet boat. The total cost of the trip was about $7.40—or 740¢. How many miles long was this trip? You can divide 740 by 4 to find the answer.

$$
\begin{array}{r}
185 \\
4{\overline{\smash{\big)}\,740}} \\
-4 \\
\hline
34 \\
-32 \\
\hline
20 \\
-20 \\
\hline
0
\end{array}
$$

The trip was 185 miles long.

This 1834 poster advertises a trip by packet boat and train from Albany to Utica, Rochester, or Buffalo. Passengers could take a train from Albany to Schenectady. At Schenectady, they could board a packet boat that would take them to Utica, Rochester, or Buffalo.

▶

1834.
PACKET BOAT,

AND
Rail-Road

ARRANGEMENT.

A Packet BOAT will leave Schenectady Daily, for Utica, Rochester and Buffalo, at

Half past 10 o'clock A. M., and

Half past 6 o'clock P. M.

PASSENGERS for the PACKETS will leave Albany by the CARS at 9 A. M. and 5 P. M.

These are the only Cars that run to the Packets.

By this arrangement there is no delay, as the Packets will leave Schenectady immediately after the arrival of the Cars.

The Erie Canal Today

Today the Erie Canal is used only for **recreational** purposes. Many people enjoy using their personal boats to travel up and down the canal. Others enjoy bicycling or hiking on the 300 miles of trails along the canal. Visitors can stop at any of the restaurants, hotels, and shops that are found along the canal.

If you want to experience what it was like to travel on the canal in the early days, you can take a ride on a packet boat. You can also visit Erie Canal Village in Rome, New York, to learn more about life along the canal in the 1800s.

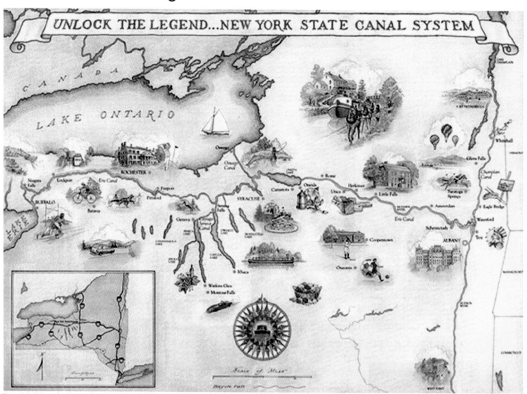

UNLOCK THE LEGEND...NEW YORK STATE CANAL SYSTEM

Glossary

aqueduct (AA-kwuh-duhkt) A bridge used to carry flowing water across a river, road, or valley.

cabin (KA-buhn) A large, enclosed room below the deck on a boat.

ceremony (SAIR-uh-moh-nee) An event to honor the importance of something, often with music and speeches.

challenge (CHA-luhnj) A task or problem that requires a lot of thought to solve.

engineer (en-juh-NEAR) A person with the knowledge and skills needed to build things like roads, dams, canals, bridges, and buildings.

packet boat (PA-kuht BOHT) A boat used to carry people on canals. It is long and wide, with a flat bottom and a cabin where people can sit.

passenger (PA-suhn-juhr) Someone who travels in a car, truck, or bus, or on a boat, plane, or train.

recreational (reh-kree-AY-shuh-nuhl) Done for pleasure or entertainment.

surveyor (suhr-VAY-uhr) Someone who measures land. Land must be measured before it can be bought or sold.

toll (TOHL) A fee paid in order to be able to use something, like a road, a bridge, or a canal.

Index